Aloha O Kalapana

Going fishing for ʻōpelu at Kaimū Bay, circa 1900.
Courtesy of Lyman House Memorial Museum.

FOLLOWING SPREAD:
Kaimū Bay and the world-renowned Black Sand Beach in 1990, just before they were engulfed by lava.

O Kalapana

Photography by Dorian Weisel

Text by Frankie Stapleton

Bishop Museum Press
Honolulu
1992

BISHOP MUSEUM PRESS

Post Office Box 19000-A
Honolulu, Hawai'i 96817

W. Donald Duckworth, director

Note: Unless otherwise indicated, photographs on these pages are by Dorian Weisel and are from his personal collection. Weisel holds copyright to his photographs that appear in this volume.

©1992 by Bishop Museum

All rights reserved

Library of Congress Catalogue Card No. 91-078237
ISBN 0-930-897-69-2

Designed by Gerard A. Valerio, Bookmark Studio

Composed in Bembo and Centaur by the
Maryland Composition Co., Glen Burnie, Maryland

Printed by the Toppan Printing Co., Japan

PRECEDING SPREAD:

Kaimū Bay and the Black Sand Beach, August 1990

Since antiquity, Kalapana has been known for its beautiful groves of coconuts. Dowager Queen Emma, widow of Kamehameha IV who ruled in the mid-1800s, commemorated the happy days she spent in the area by tending a coconut tree so that it grew in a reclining manner, one of the "niu moe o Kalapana."

Joseph V. Veriato, Jr., (at right) with Deputy Civil Defense Director Bruce Butts. This book is dedicated to "Junior," whose heart could take no more and gave out during the 1990 eruption, and to all the other caring Hawai'i Island citizens who labored long, often nerve-wracking hours on behalf of the residents of Kalapana, Kaimū, and Kapa'ahu, especially civil defense's Harry Kim, Bruce Butts, and Lanny Nakano. Most of all, this account, which represents mere samples of the rich history of the Kalapana area and its people, is dedicated to those born and laid to rest there, as well as to the future residents of Kalapana.

Contents

Foreword   x

Kalapana: A Paradise Lost   3

CHAPTER 1: The Warning   7

CHAPTER 2: Geology, the Hawaiians, and Pele   15

CHAPTER 3: A Polynesian Dynasty   25

CHAPTER 4: Kalapana's First Tourists   29

CHAPTER 5: A Tenuous Hold on the Land   35

CHAPTER 6: An Observatory and Park on Kīlauea   43

CHAPTER 7: From Puʻu ʻŌʻō to Kūpaianaha   53

CHAPTER 8: The Beginning of the End   67

CHAPTER 9: The Onslaught   87

CHAPTER 10: The Heart of Kalapana   99

CHAPTER 11: The Finality   127

CHAPTER 12: The Legacy   147

Bibliography   152

Foreword

Love is an amazing thing. Not "fallen" into other than naturally and surely by no design but its own.

The Earth knows no sleep. Day and night, spring and fall. Change! Always change. To live in Hawai'i is to know this change, to be part of it, to thrive on it, to allow it to inspire your passions.

Kaimū . . . the seduction of the rustling palms. The rhythm of the sea. That magic temperature when you forget your body and float on the breeze . . .

Take a deep breath . . . listen . . . really listen to the sea. Never forget these palms, swaying in that sweet breeze . . . oh, how they seduce me . . . to float across volumes of memories, miles of images—so many of that brief moment when I caught a glimpse of Kalapana, before her timelessness was pierced, before she went the way she came.

I first saw Kīlauea erupt in 1979. But it was that incredible beacon in the sky every month in '83 and '84 that worked its magic on me. By 1985 I had moved to Volcano and the mountain's eruptive cycle became a part of my own.

There was a moment, maybe a few, when the intensity of the event became so overwhelming . . . but you go home, eventually you must. After days of having rocks rain down and collect under my collar and in my boots I *wanted* to go home. Back to civilization; what would I do without a hot shower?

Kalapana, May 1990 . . . disaster would be better if it came unannounced, did its business and was gone. This one didn't happen that way.

I came to town with the lava. Because it was so slow in doing what it must, I met an incredible number of people . . . It was mine to meet someone one day, witness their home incinerated the next and they would be scattered to the wind on the third. The overwhelming loss of community.

I never knew Kalapana, her greater peace, the celebration of the 'āina that bonded her people together. I met them and am full of their stories, of how they came, and how they loved, of the people, the land, every sentence with the word Kalapana in it . . .

Puna Canoe Club's longhouse, where racing canoes were kept.

Every day a new friend, another's home a part of the evening glow, the growing lava field. The landscape knows change. Kalapana is being destroyed . . . I am falling in love.

I would steal away to some corner, take off my boots and allow the peace of the *keiki* pond to soothe me. I would fantasize myself a life along this stretch of coast.

Kaimū . . . for the longest time seemed so far removed from the threat . . . a cup of coffee at Chings' . . . a busload of tourists would waddle onto the beach, cameras, smiles, a few grains of black sand wrapped into a handkerchief and back to the bus . . . ah, the sweet smells of reality, the sights and sounds. Everything except what you had to face just around the corner . . .

Kaimū . . . it was here that I would seek refuge in the night. Here I would go to heal after another yard full of fruit trees "planted when we first cleared the land" was overrun . . .

I realized I was living through the destruction of something I loved. Something very dear to me was passing.

With no claim other than that of a lover, I too was having her ripped from my heart . . .

Remember what it felt like to drive down 130 and turn to Kaimū? *Remember* that wonderful breeze? The contrast between the ocean, palms and black sand all so wonderfully set off by the pure white of the ocean's foam at the water's edge . . .

Dorian Weisel

Aerial view of the Kalapana coast looking to the northeast, with the hub of Kalapana in the left center and Kaimū Bay in the background.

Aloha O Kalapana

"There are a lot of people who enjoyed living in Kalapana, Kaimū, and Kapaʻahu in a peace—in a way of life that was a little different and a lot different from many of ours. They enjoyed their lifestyles. Their goals may have been different from what some in the western or highly industrialized world would aspire to, but Kalapana was one of the few, if not the last place, where a certain way of life of old was still being maintained. And they could do it because they had their land. Their land was abutted by the ocean, not by expensive resort developments. And this is what they lost."

—Harry Kim, Hawaiʻi County
Civil Defense administrator

Kalapana: A Paradise Lost

Palms damaged during the 1975 earthquake at Kalapana are silhouetted in this early morning view, taken in the spring of 1990, of the brackish ponds behind the Star of the Sea Painted Church.

Kalapana today is a state of mind: a memory and a vision. The once-legendary seaside community—cherished since ancient times for its beauty, its graceful coconut palms and sweet-smelling *hala* trees, its exotic black sand beaches, and its simple, natural lifestyle—was entombed in a thick blanket of lava in 1990.

Kalapana and the neighboring communities of Kaimū and Kapaʻahu were three historic settlements located along the coastal boundary of Kīlauea Volcano, one of five volcanoes making up the island of Hawaiʻi. The traditional home to generations of native islanders who trace their heritage from prehistoric times, the isolated Puna communities—known singularly although unofficially as Kalapana—had resisted the impact of westernization longer than any other district in the four major Hawaiian Islands.

In an attempt to preserve the eroding native lifestyle centered around fishing and farming, Congress in 1938 authorized a Kalapana Extension to Hawaiʻi Volcanoes National Park, granting protected status to the Hawaiians living along the Puna coast. In 1981, the federal government completed construction of a modern highway linking Kīlauea's summit with the coast, a thoroughfare that brought thousands of visitors through the national park and downslope into the quiet rural villages. Outsiders began moving into the picturesque area as low-cost subdivisions were carved into the volcanic landscape, and by the time Kīlauea began its extended East Rift Zone eruption in 1983, Kalapana had become two communities: one of native Hawaiians who still lived on their ancestral lands and the other of newcomers, including many young people and retirees seeking an affordable life in paradise.

Today, for all but a few natives of the land, memories are all that remain of Kalapana, as Kīlauea Volcano has erased humanity's inroads into the rugged, hauntingly beautiful domain of the traditional volcano goddess, Pele. How this unique cluster of communities faced conquest by Nature is a testament to the enduring spirit of the people of Kalapana, Kaimū, and Kapaʻahu.

"Civilization exists by geological consent,
subject to change without notice . . ."

Will Durant

CHAPTER I
The Warning

The idyllic calm of life in Kalapana shattered one pristine day in September 1977. The serenity of the picturesque community along the eastern coast of the island of Hawai'i in the District of Puna proved illusory as civil defense officials called the residents together to alert them: The brooding Kīlauea Volcano in whose shadow Kalapana and the neighboring communities slumbered had erupted along its East Rift Zone.

The isolated settlements—home to Hawaiian descendants of those who had populated Kīlauea's slopes and coastal plains for centuries, a smattering of Mainland retirees, and several itinerant surfers—were in the path of an encroaching lava flow moving downslope toward the sea. In 1977, Kalapana, Kaimū, and Kapa'ahu represented the largest concentration of Hawaiians in the four major islands practicing what remained of the native lifestyle. The cultural and geologic uniqueness of the community—from Waha'ula Heiau to Kaimū Bay—had prompted efforts by federal authorities, beginning in the 1930s, to preserve and protect it.

But on that fateful September day, government authorities were warning the residents that the lava on a march to the sea was aimed directly at the heart of Kalapana, and nothing could stop it.

Less than a week later, despite the dire prediction, the lava flow ground to a halt three-quarters of a mile above Kalapana, sparing the coastal hamlet. The impetuous, quixotic forces of Nature, which have ruled all life along this rocky shore from ancient times to the present, had prevailed.

But the black finger of lava that remained frozen above the town center was visible warning of the potential for disaster. And in spite of the recent volcanism, there was an influx of new residents over the next few years.

The village of Kalapana occupied a long ground depression known as a graben, which was partially flanked by an area of higher ground called a horst. In this 1988 view, the horst is seen jutting into the ocean before the coastline turns back to form Kaimū Bay (background).

The outpouring from Kīlauea's East Rift Zone began on January 3, 1983, from fissures near Nāpau Crater in Hawai'i Volcanoes National Park, about eight miles from the volcano's summit. After settling in at a site farther downrift, near the park's boundary, the eruption proceeded throughout the 1980s, creating the massive flow field seen descending the pali, *the mountain ridge. The longest-running eruption of Kīlauea's rift zone in recorded history, the volcano produced lava flows that threatened, and finally, in 1990, demolished Kalapana.*

When Kīlauea's quivering East Rift Zone next erupted, on January 3, 1983, spewing acrid smoke and incandescent fountains of lava, no one, not even the geologists at the Hawaiian Volcano Observatory, could predict the eruption would last through the decade. Or that it would destroy the last remnants of a culture that had coexisted with the volcano for more than a millennium.

But in a land where cataclysmic events seemingly always have been a part of life, as the eruption wore on throughout the 1980s and eventually demolished Kalapana in 1990, the realities of living atop a still-growing volcano taught a hard lesson to the people living in the lava's path.

A ground view of the Kūpaianaha vent

Streams of lava ate through the thick forests on the slopes above Kalapana in the late 1980s as the eruption continued, eventually forming lava tubes that carried the molten material miles downslope.

CHAPTER 2

Geology, the Hawaiians, and Pele

The awesome forces of the volcano, from those witnessed by night visitors to the lava lake at Kūpaianaha to the formation of new black sand beaches where lava meets the sea, have long mesmerized and inspired mankind.

Earth is a living, breathing entity in a ceaseless process of eruption, subduction, creation, destruction, and regeneration. The ancient Hawaiians knew this to be so a thousand years ago; it took the civilized world a bit longer to discover. It was more than a decade after Captain James Cook's voyages of exploration in the Pacific and his discovery of Hawai'i for the western world that the father of modern geology, a Scotsman named James Hutton, proposed in his *Theory of the Earth* that our planet was much older than the six thousand years then believed. Hutton saw that geologic time was much longer, that Earth was a perpetual force, ever changing through growth, decay, and revitalization.

The Hawaiians personified and deified Earth's awesome powers. Their gods of Nature dominated all aspects of daily living, indeed set the rules by which they lived and tempered their concept of the universe and of life. These concepts were passed from generation to generation through associated practices, legends, songs, and chants. And the undulating, venting activity of the volcano was inspiration, according to some experts, for the particular style of *hula* developed in Hawai'i.

Early scholars were intrigued by analogies between Hawaiian myths and scientific theories. The resemblance of the *Kumulipo,* the Hawaiian creation story, to the scientific theory of evolution is one example. The *Kumulipo,* a long genealogical chant composed about 1700, tells first of a period of darkness during which lower forms of life gave way to pairs of sea and land plants, fishes and birds, creeping reptiles and mammals—forms that were subsequently revered as gods. A transformation from darkness to light followed, and the coming of the great gods and mankind signified the world of humanity from which Hawaiian family histories evolved.

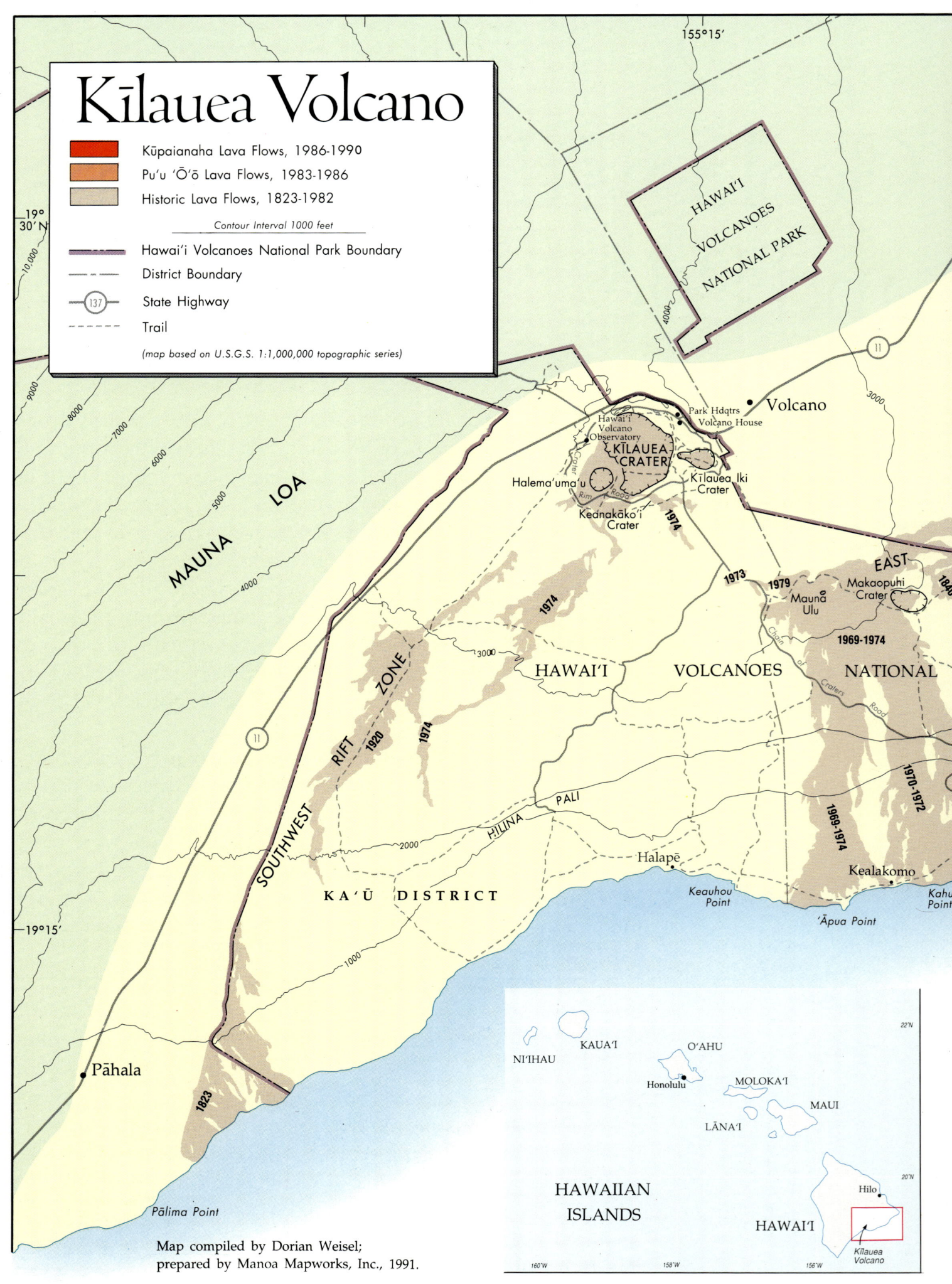

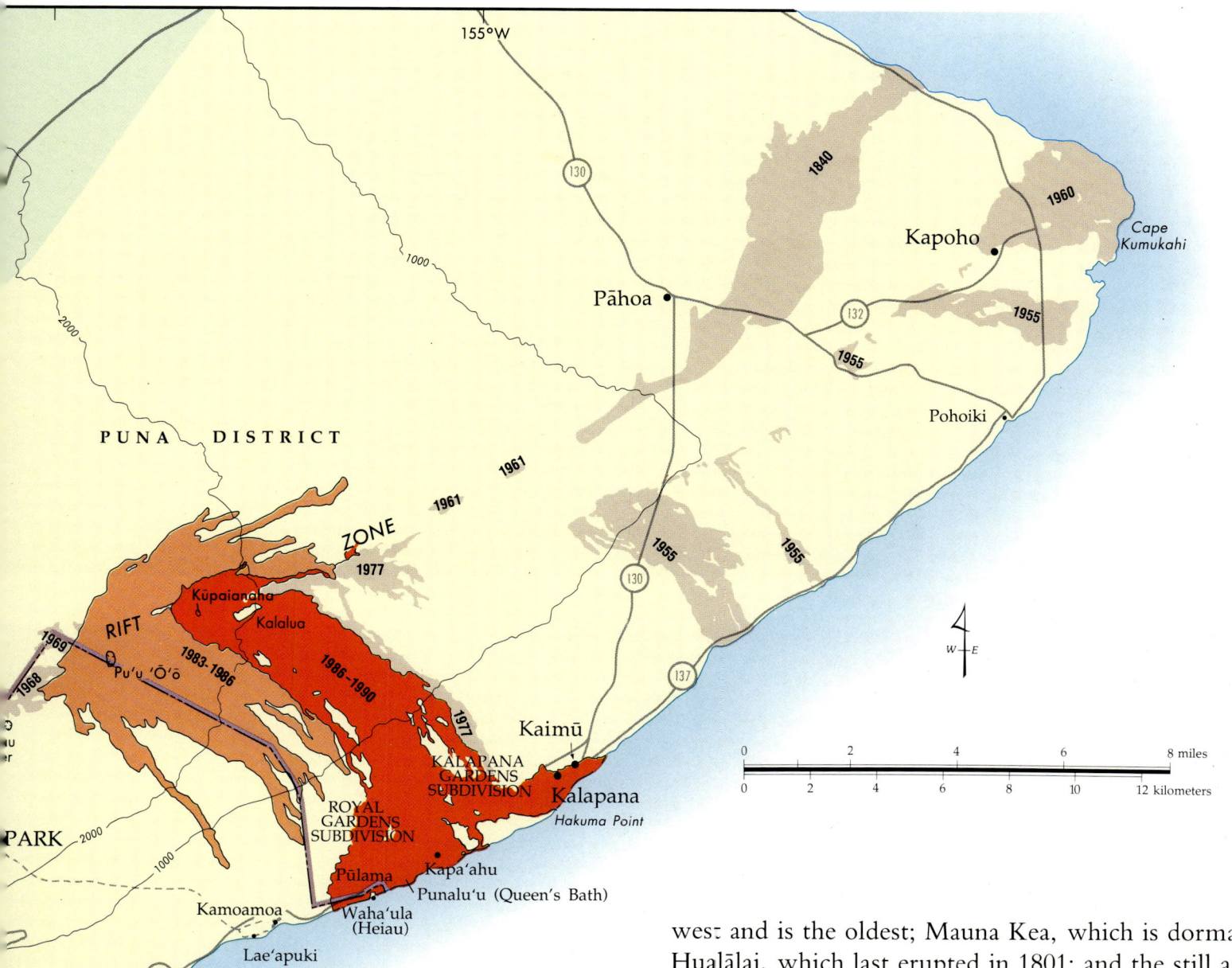

Hawai'i Island—approximately 700,000 years of age and still growing—is the youngest in a series of volcanic islands that stretches sixteen hundred miles from the southeast to the northwest. A permanent hotspot beneath the central Pacific Ocean produced the volcanoes that erupted through the migrating Pacific Plate to create the Hawaiian Archipelago. The archipelago extends from the seamount Lōihi twenty miles off the Puna coast to the Northwestern Hawaiian Islands and ends at Kure Atoll, a coral island approximately thirty-seven million years old.

Five volcanoes form the land mass of the Big Island of Hawai'i: Kohala, which occupies the northwest and is the oldest; Mauna Kea, which is dormant; Hualālai, which last erupted in 1801; and the still active Mauna Loa and Kīlauea.

Since 1955, Kīlauea's East Rift Zone overlooking Kalapana, Kaimū, and Kapa'ahu has been the primary focus of that volcano's activity. This zone extends some thirty-five miles above ground, from Keanakakoi Crater at the summit northeast to Cape Kumukahi, and another forty miles under the Pacific Ocean.

Kīlauea's destructive forces are not limited to outpourings of lava. Buttressed by the massive Mauna Loa on its north and west sides, the still-growing volcano is only able to move outward along its southern flank, toward the open ocean. In another of Nature's constant cycles of growth and destruction, the seismic activity created by the upwelling of magma from deep within the Earth causes pieces of the coastal plain to break off and subside into the ocean. Historic villages from Honu'apo and Punalu'u in Ka'ū to Kalapana and Pohoiki along the Puna coast have been subjected to gigantic ocean waves known as tsunamis and to subsidence, both triggered by earthquakes.

Kīlauea's volcanism, fed by a plume of magma rising from the Earth's plastic mantle, about forty miles below the ocean floor, was centered in the summit's caldera in the nineteenth and early twentieth centuries. A boiling lava lake agitated almost continuously within the volcano's gigantic sunken pit (seen behind Kīlauea Iki in the foreground at left) until the lake suddenly drained in 1924 and a series of steam explosions created Halema'uma'u Crater. Since that time, eruptive activity at the summit has been limited to short, sporadic outbursts.

As magma forces its way up from the semi-molten mantle, the brittle crust of the Earth's outer layer must give way for the new material. The resulting seismicity, or ground vibrations, range from deep-seated earthquakes, generated at depths of more than forty miles in the mantle, to crustal adjustments in the shallower depths less than ten miles down. Kīlauea's surface expands and deflates with the infusion and expulsion of molten material. And as the world's most active volcano continues to grow, something has to give.

On November 29, 1975, a 7.2 earthquake beneath Kalapana resulted in subsidence of low-lying portions of the coast along much of Kīlauea's southern flank, with accompanying tsunamis that drowned two people on a Boy Scout outing at Halapē. Scientists theorize that the earthquake relieved internal stresses that had been building up in the volcano's flank for decades and created new openings for the ongoing subterranean infusion of magma into the East Rift Zone.

But the most enduring of the Hawaiian legends and one that is still winning converts today is that of Pele, the fiery goddess of the volcano who left a distant land in search of a suitable home to contain her volatile temperament. The Pele myths, the subject of countless chants and *hula,* tell the story of the restless goddess's migration from the land of Polapola in Kahiki (Bora Bora in the Hawaiian language) to the northwestern shoals and down the Hawaiian Island chain, traveling southeast from island to island, digging with her ʻōʻō stick to find a home for herself and her brothers and sisters. But her jealous older sister, Na-maka-o-kahaʻi, a sea goddess whose lover she had seduced, pursued her in anger, using her overwhelming tides to drive Pele on. When Pele arrived at Hawaiʻi, the youngest of the islands, on the southeastern end of the chain, she dug deeply and found within Halemaʻumaʻu the perfect refuge for her smoldering nature, and it is there, in spirit form, that she settled.

There are many variations of Pele legends, but the Hawaiians of old believed that the boiling caldron of Kīlauea was the domain of Pele and her family. Their arrival in an area, the ancient people said, was traditionally announced by convulsive quaking of the Earth, flashes of blue lightning, and a tremendous roar of thunder. Sometimes Pele traveled underground from her home in the crater to locations downslope, sending lava flows to take vengeance on those who would dare offend her.

The Legend of Keliikuku

"According to common tradition the district of Puna was, until two centuries ago, a magnificent country, possessing a sandy soil it is true, but one very favorable to vegetation, and with smooth even roads. The Hawaiians of our day hold a tradition from their ancestors, that their great grandparents beheld the advent of volcanic floods in Puna. Here in brief is the tradition as it was told by the islanders in Jules Remy's *Contributions of a Venerable Savage to the Ancient History of the Hawaiian Islands.*

"This high chief who reigned in Puna journeyed to the island of Oʻahu. There he met a prophet of Kauaʻi, named Kaneakalau, who asked him who he was. 'I am,' replied the chief, 'Keliikuku of Puna.'

"The prophet then asked him what sort of country he possessed. The chief said: 'My country is charming, everything is found there in abundance, everywhere are sandy plains which produce marvellously.' 'Alas!' replied the prophet, 'Go, return to your beautiful country. You will find it overthrown, abominable.

Pele has made of it a heap of ruins; the trees of the mountains have descended towards the sea, the 'ōhi'a and pandanus are on the shore. Your country is no longer habitable.'

"The chief made [an] answer: 'Prophet of evil, if what you now tell me is true you shall live; but if, when I return to my country, I prove the falsity of your predictions, I will come back on purpose and you shall die by my hand.'

"Unable, in spite of his incredulity, to forget this terrible prophecy, Keliikuku set sail for Hawai'i. He reached Hāmākua and, landing, travelled home by short stages. From the heights of Hilo at the village of Makahanaloa, he beheld in the distance all his province overwhelmed in chaotic ruin, a prey to fire and smoke. In despair, the unfortunate chief hung himself on the very spot where he first discovered this sad spectacle."[1]

[1] Source: Allen, Melinda Sue. 1979. "The Kalapana Extension in the 1800s: A research of the historical records." National Park Service, Hawai'i Volcanoes National Park. Typescript. Originally published in *Contributions of a Venerable Savage to the Ancient History of the Hawaiian Islands,* by Jules Remy, 1868, privately printed and translated from the French by William T. Brigham.

The ancient Hawaiians saw their universe as controlled by spirits, personified in the natural world around them, which they feared and revered and to which they paid homage and supplication.

OPPOSITE PAGE:
Puna's forested slopes fall prey to the fire and smoke of a 1990 lava flow from Kīlauea's vent at Kūpaianaha.

Ten traditional partitions of land covering more than forty-nine thousand acres were added to Hawai'i Volcanoes National Park through the Kalapana Extension Act of June 20, 1938: From the tip of Keauhou in Ka'ū and 'Āpua at the southernmost point, through Kahue, Kealakomo, Pānau-nui, Pānau-iki, Lae'apuki, Kamoamoa, and portions of Pūlama and Poupou. The extension included the ruins of Waha'ula Heiau, the village of Kealakomo, and the Pu'uloa petroglyph fields. The Territory of Hawai'i acted as agent for the federal government in obtaining the lands, many of which, as crown properties and governmental lands, were already under territorial control.

As a condition of the 1938 act which took possession of their ancestral lands, Congress granted native Hawaiians in the Kalapana area protected status to conduct religious practices as well as to fish, hunt, and gather plants in the national park. Today, some of the area's native Hawaiians are also employed by the national park as resource specialists, interpretive guides, and park rangers.

CHAPTER 3

A Polynesian Dynasty

Nineteen hundred years or more have passed since the arrival of the first Polynesians to Hawai'i, and Nature has left scant evidence of those whose spirits and passions brought them from shores thousands of miles distant to settle new lands. One of the few ancient sites to survive the onslaught of development as well as inundation by lava and the sea is Waha'ula Heiau, located in Hawai'i Volcanoes National Park four miles south of Kaimū.

This most sacred temple was built upon arrival on Puna's shores of the priest Pā'ao and his people from the ancestral homeland of Kahiki about A.D. 1250. Legends tell of Pā'ao, a powerful kingmaker and one of the great sea-voyaging Polynesians who colonized the Pacific in the thirteenth and fourteenth centuries, as introducing new chiefly bloodlines, edible plants, and revolutionary practices.

Together with Pili-ka'aiea, a chief associated with the Kahiki connection, Pā'ao ruled Hawai'i. He established new bloodlines for the *ali'i* (chiefs) and *kāhuna* (priests), lineages that survived five hundred years and included Kamehameha I and his high priest Hewahewa approximately twenty-one generations later. Pā'ao saw to it that the existing priesthood was squashed while he imposed severe new rituals, including human sacrifice, that gave life-and-death power to the chiefs and the keepers of the temples. The *kāhuna* introduced by Pā'ao oversaw a rigid system of *kapu* or taboos, rites that ruled every aspect of life for commoner and chief alike.

Waha'ula, meaning red mouth, was a *luakini heiau,* a temple of the highest order with the strictest of *kapu,* of such sacred power that even the smoke rising from its sacrificial altars meant death to anyone who passed within its shadow. Originally built to house Pā'ao's god within a *kīpuka* of *'a'ā* lava, the *heiau* was

constructed in the shape of a parallelogram, with walls more than 10 feet in height, 132 feet long, and 72 feet wide. A series of terraces led to platforms within the interior where sacrifices were performed. Because of its proven power, it was rededicated to the god Kū by rulers who came generations after Pā'ao, including Imaikalani, Kalani'ōpu'u, (the elderly ruling chief of Hawai'i who greeted Captain Cook in Kealakekua Bay in 1779), and Kamehameha I. A commoner could not even approach this *heiau,* as access was restricted to royalty and the priests of the temple.

The grounds in the vicinity of Waha'ula also had special uses. Nearby was a testing area for warriors in training, where young men practiced spear throwing by hurling a ten-pound stone to a wall seventy feet away. A young warrior's strength was measured by the distance the stone bounced back toward him. Not far from Waha'ula, within a couple of miles, there was a *pu'uhonua,* a place of refuge; stepping within its boundary meant immediate sanctuary for wrongdoers, injured soldiers or women and children fleeing a conquering chief, or a wife escaping a husband's wrath.

Religious rites, including human sacrifices, brought to Hawai'i in the thirteenth century by the priest Pā'ao from Kahiki, were first practiced in Pūlama, at Waha'ula Heiau (foreground). After building Waha'ula and establishing his religious order in Puna, Pā'ao constructed another sacrificial temple in Kohala and the strict new rituals subsequently spread through the Hawaiian Islands.

Following the acquisition of Pūlama as part of the Kalapana Extension lands, the National Park Service built the Waha'ula Visitor Center for displays and interpretive presentations. In the aerial view (left) from the sea, the visitor center can be seen just beyond the ruins of the walled temple in the foreground. The metal skeleton of the center was all that remained after a lava flow incinerated the building on June 22, 1989. Despite flows encroaching on all sides, the oldest section of the Waha'ula Heiau, built on *'a'ā* created fifteen hundred to ten thousand years ago, was left standing.

CHAPTER 4
Kalapana's First Tourists

The inland pool at Punalu'u, known popularly as Queen's Bath in modern times, is associated with several ancient legends. According to one account, a heiau was built near the pool to commemorate the death of Punalu'u, the shark-man.

In 1823, the Hawaiian guide Mauae approached his birthplace at Kaimū escorting an English minister, the Reverend William Ellis. As part of the missionary effort that was just taking root in the islands, Ellis, who had served as a missionary for six years in Tahiti and adapted quickly to the Hawaiian language, was dispatched on an exploration of the island of Hawai'i. Landing at Kailua on the western shore of Hawai'i in June 1823, Ellis and three American missionaries set out on foot to explore the island and its people, with the purpose of establishing future missions. In the course of their two-month tour of Hawai'i, Ellis and his party became the first foreigners to witness an eruption of Kīlauea at its summit as they traveled through what is now part of the Hawai'i Volcanoes National Park to the villages of Kalapana and Kaimū.

Much in the lifestyle practiced in Hawai'i as witnessed by the touring missionaries had existed unchanged for hundreds of years. In their explorations, the island's first tourists not only viewed "the fiery surge and flaming billows" of Kīlauea, but numerous and various settlements, cultivated plantations, aquacultural ponds, lava plains used for salt production and the curing of fish, temples large and small, and the practices of a people attuned to the rhythms of Nature.

As Ellis entered the village of Kaimū led by the native guide, multitudes of young people and children turned out, chanting Mauae's name, the name of his parents and the story of his birth, and the outstanding events in his family's history, weeping for joy at the sight of their beloved one.

The early western explorers such as Ellis found a highly developed Hawaiian culture, with values and expertise in various skills passed from one generation to another through *kāhuna* or masters. There was no written language. The oral tradition was all-important as the ancient Hawaiians preserved their legends and mythology, traditions, customs and beliefs, and family and cultural histories by word of mouth, through songs or chants.

Kalapana's Congregational church dominates the view from the mountain side of the seaside settlement.
Photo by the Reverend Albert S. Baker, date unknown. Courtesy of Lyman House Memorial Museum and the Mission Houses Museum Library

Palm leaves cover outrigger canoes in the foreground of this view of Kalapana, circa 1890.
Collection of Bishop Museum.

Knowledge of such matters was closely guarded and shared only with selected individuals, usually family members, who were deemed worthy of handling such valued information. The ancient legends and chants, including genealogical chants, were their method of retaining and passing on their accumulated knowledge. Until the introduction of written language with the arrival of the foreigners, Hawaiians utilized oral tradition to record their history.

In the four years since the death of Kamehameha I in 1819, Hawai'i had experienced the royally dictated overthrow of traditional nativistic beliefs, spurred on in the 1820s by the conversion to Christianity of several of Kamehameha's wives. Thus, the missionaries gained lands for churches and schools and were permitted to preach to the people about their god, Jehovah.

On its descent from Kīlauea en route to Kalapana, the Ellis party found a "substantial" population in Puna, with the greatest concentrations of people inhabiting settlements near the coast and smaller clusters inland. Fishing—both in the deep ocean and along the shore where limpets and other shellfish, sea urchins, and seaweed were gathered—provided the mainstay of their subsistence. Dried fish and salt were their articles of commerce, to be traded with other commoners and given in payment to their chiefs for rights to use the land. Groves of coconut trees, taro cultivated in terraces, sweet potatoes grown in sheltered pockets in the lava, and plantings of sugar cane provided other sustenance, as did pig, fowl, and dog.

Their *hale,* houses, were simple one-room stone and thatch structures, often built for a singular purpose: an eating house; a sleeping house for shelter at night or for protection from bad weather; a storehouse for keeping supplies, canoes, or fishing implements; or a workhouse for weaving or making *kapa* cloth. The commonfolk, *maka'āinana,* spent the majority of their time outdoors, doing most of their cooking, eating, craft work, and playing under the open sky.

After witnessing the affection and joy showered on their young guide and enjoying the hospitality and bounty provided by his family, Ellis and the other visiting ministers preached and sang hymns for the villagers under the shade of the trees. The religious men were impressed by the people and their industriousness as well as the beauty and cultivation of the neighborhood. When a tour of the area indicated a population of about two thousand, Ellis determined Kaimū to be a good site for a new mission. That mission was later named the Kalapana Mauna Kea Congregational Church.

Villagers turn out to watch the launching of a canoe, circa 1900. Hawaiian fishermen traditionally fished cooperatively and shared their catch at the end of the day.
Courtesy of Lyman House Memorial Museum.

"Kīlauea Night Scene" by Titian Ramsey Peale, 1842. Collection of Bishop Museum.

"The fires of Kilauea, darting their fierce light athwart the midnight gloom, unfolded a sight terrible and sublime beyond all we had yet seen. The agitated mass of liquid lava, like a flood of melted metal, raged with tumultuous whirl. The lively flame that danced over its undulating surface, tinged with sulfurous blue, or glowing with mineral red, cast a broad glare of dazzling light on the indented sides of the insulated craters, whose roaring mouths, amidst rising flames and eddying streams of fire, shot up at frequent intervals, with very loud detonations, spherical masses of fusing lava, or bright ignited stones. The dark bold outline of the perpendicular and jutting rocks around formed a stark contrast with the luminous lake below. whose vivid rays thrown on the rugged promontories and reflected by the overhanging clouds, combined to complete the awful grandeur of the imposing scene."

The first written description of Kīlauea at night, as recorded in 1823 by the Reverend William Ellis.

CHAPTER 5

A Tenuous Hold on the Land

The people of old Hawai'i did not own and dwell on fixed pieces of land. In their hierarchical yet cooperative society, rights to use lands were granted by the chief to an elite ruling class, which in turn gave rights to commoners. Communities typically utilized the full range of their *ahupua'a,* the traditional division of land that extended from the mountain uplands to the sea. If the fires of Pele overran an area and turned the land to stone, or the ocean rose up to swallow a canoe-landing area, the survivors simply, as a matter of course, relocated within the *ahupua'a,* either at the coast or to the misty uplands. Their flexible lifestyle reflected a recognition of the changeable nature of life in the land of volcanoes.

Archaeological finds and historical accounts reveal evidence along the Puna-Ka'ū coastline of numerous fishing villages. Some, at 'Āpua, Keauhou, Kealakomo, Kahue, Pānau, Lae'apuki, Kamoamoa, Ka'ili'ili, and Poupou-Kauka, were still in existence as late as the 1850s. For the people at Kealakomo and, later, at Kamoamoa, salt, for the preservation of foods and preparation of the dead, provided a product for exchange. The uplands held *kalo* patches, often grown in groves of *hapu'u,* and forests where birds were caught for plumage to decorate the cloaks and helmets of the chiefs. Medicinal herbs were found both in the mountains and near the shore.

The Hawaiians' communal lifestyle relied on the exchange of goods and the land tenure system in which ruling chiefs controlled the lands of an *ahupua'a.* Money in exchange for goods and services came to Hilo in the 1840s by introduction through Captain Charles Wilkes of the United States Exploring Expedition. With the Great Mahele of 1848, Hawai'i adopted the foreign

system of private ownership of land. Many in Puna's coastal communities, unfamiliar with the new monetary system and their rights as native tenants, soon thereafter lost claim to their ancestral lands. These people, living away from the centers of commerce, practiced a subsistence lifestyle, earning their livelihood from the land and the sea. This lifestyle continued into the early decades of the twentieth century.

Living with the turbulence of Nature's yin-yang forces of creation and destruction was very much a part of life in Hawai'i. In the spring of 1868, the Ka'ū and Puna districts were ravaged by a series of cataclysmic events for more than a week, including thousands of earthquakes, a catastrophic tidal wave, and volcanic eruptions of both Kīlauea and Mauna Loa. Written accounts

Aloha *has been a way of life for the people of Kalapana since the days of old. The first visitors to the area, the Reverend William Ellis and his touring group of missionaries, experienced it in 1823. These children enjoyed it in 1913. The subsistence lifestyle practiced in Kalapana into the twentieth century depended on this* aloha, *this spirit of sharing as a way of life. In Kalapana, youngsters were taught that it was not*

what one had that was important; it was the sharing of what one had that was the measure of a Hawaiian.

Worldly goods did not impress the people of Kalapana. Their riches were the simple pleasures: the bountiful ocean, the beauty of Nature, the joy of playing in the sun, of singing or talking story with others.

describe the slopes from Ka'ū to Hilo as vibrating "in an almost constant tremor," at times preventing people from standing upright and causing seasickness.

In a letter to his cousin, Frederick Lyman, son of the Hilo missionary couple David and Sarah Lyman, described events that took place in Ka'ū on the afternoon of April 2, 1868:

"I went to work outside . . . hewing some logs about 10 or 12 inches thick and 20 or so feet long, for gate posts. Two of the children were with me, sitting on some of the logs. About 4 o'clock it shook as usual, but did not stop—shook east and west, north and south, round and round, and up and down—lessen, then increase in violence. The logs we were on rolled back and forth—I called to the children to keep their feet up so as not to get caught under the logs. The stone walls were all shaken down, not one stone

While it is true the Hawaiians have existed in the shadow of Kīlauea for more than fifteen hundred years, written history over the past 175 years has witnessed the snuffing out of one village after another by the volcano's earth-force. Descendants of the families who populated Honuʻapo (above), Keauhou, ʻĀpua, Kealakomo, Kamoamoa, and the other unrecounted settlements along the Puna-Kaʻū coast are among the residents of Kapaʻahu, Kalapana, and Kaimū who were—once again—displaced by Kīlauea.

Photo by Alonzo Gartley, Collection of Bishop Museum.

Fishing and farming continued to sustain the native people in Puna into the early decades of the twentieth century. But the arrival of foreigners and the subsequent major social upheavals of the 1800s had caused serious disruptions in their lifestyle, decimating the population and undermining their spiritual and ancestral connections to the land and the natural world around them.

Photo by H. R. Hanna, circa 1905. Collection of Bishop Museum.

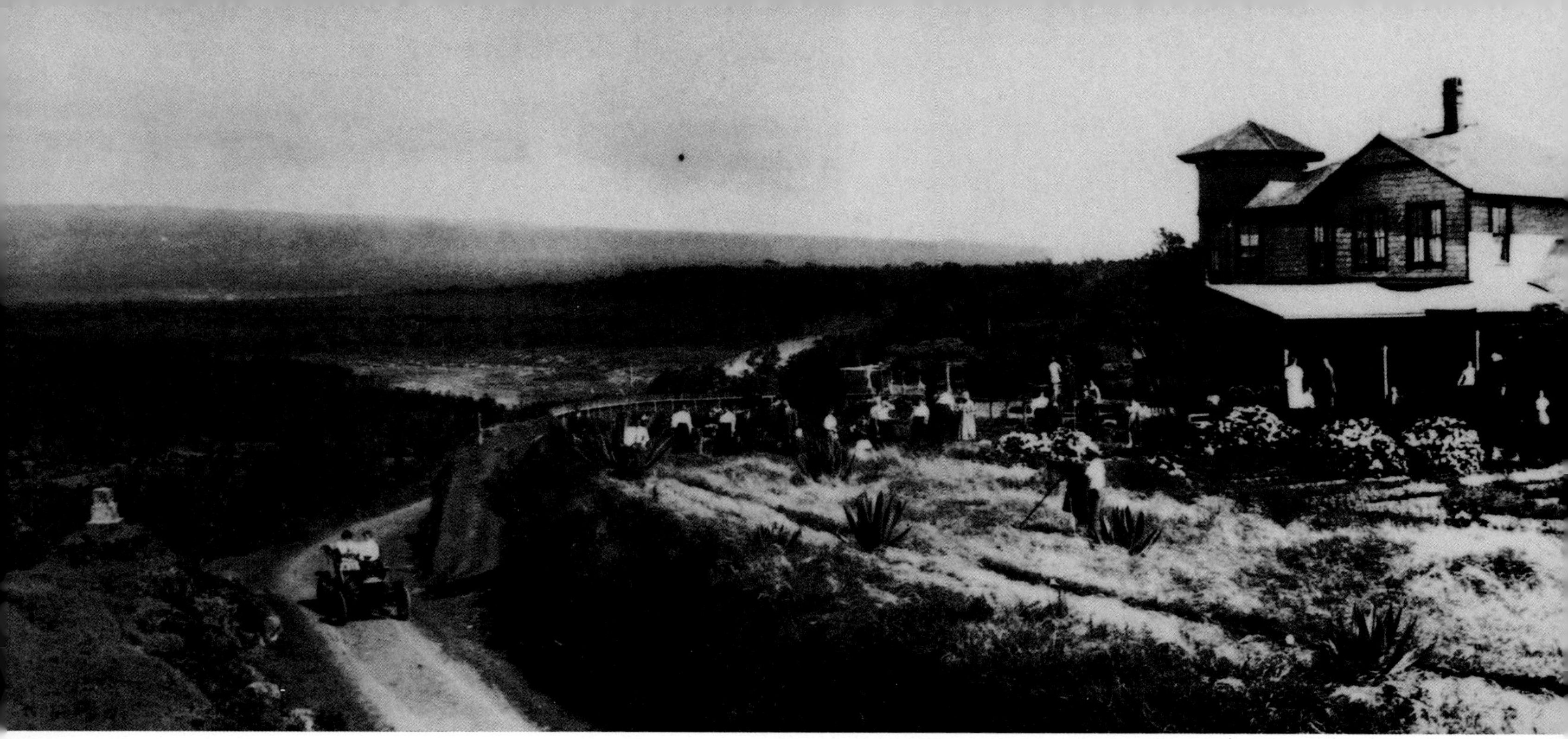

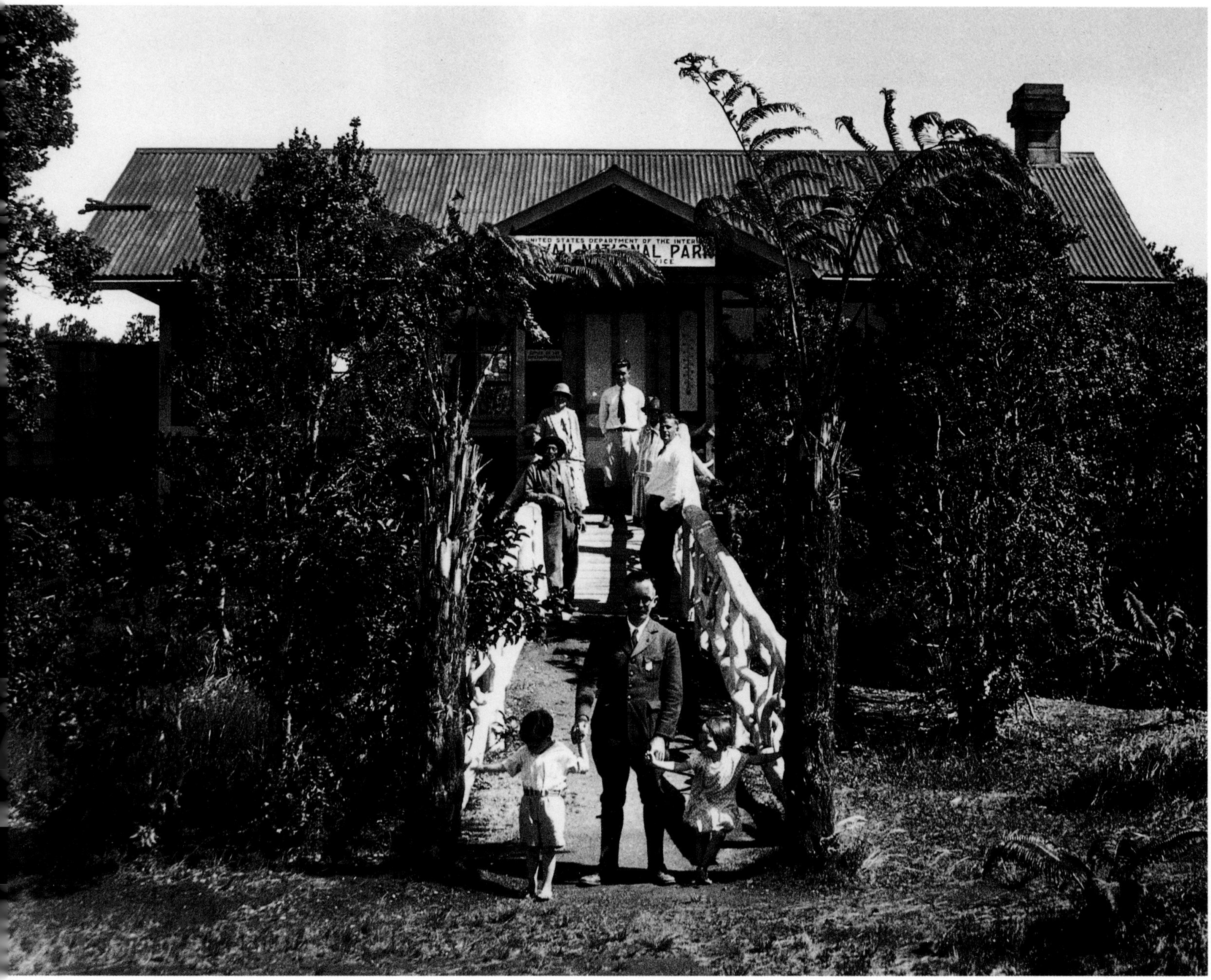
UNITED STATES DEPARTMENT OF THE

"The Puna District is the one remaining section which has least felt the coming of the white man, and should be protected to keep it as unchanged as possible."

—Everett Brumaghim, advisor to the National Park Service, writing on March 29, 1933, about the acquisition of part of Puna for Hawai'i Volcanoes National Park.

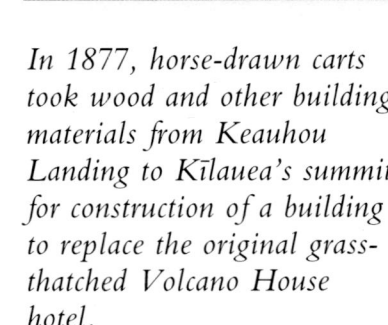

In 1877, horse-drawn carts took wood and other building materials from Keauhou Landing to Kīlauea's summit for construction of a building to replace the original grass-thatched Volcano House hotel.
Courtesy of Lyman House Memorial Museum.

reintroduced. Hawaiian Volcano Observatory founder Jaggar was then persuaded to press Congress on the issue of the park, which was being held up due to financial considerations. Under Jaggar's direction the bill was redrawn, eliminating any appeal for financial appropriations and adding to the park more than twenty-one thousand acres of the dormant volcano of Haleakalā on Maui.

Following public lectures on Kīlauea and Mauna Loa and a congressional hearing with testimony from Dr. Jaggar, delegate Kalanianaole and a representative of the Chamber of Commerce of Honolulu, Sidney Ballou, the Jaggar bill was passed by both houses of Congress with added provisions related to the acquisition of privately owned lands. It was signed into law by the president of the United States on August 1, 1916.

In 1922, Edward G. Wingate, a U.S. Geological Survey topographer who later became superintendent of Hawai'i Volcanoes National Park, was assigned to survey the Puna shoreline, which Observatory director Jaggar believed was sinking. There the surveyor found "the largest concentration of Hawaiians I have seen in the islands." The people were living much as they always had—fishing in the sea and planting *kalo,* sweet potato, and other crops inland, Wingate observed.

Westerners drawn to Hawai'i to study the natural phenomena of volcanoes found substantial communities of Hawaiians living along the Big Island's eastern shore, practicing their traditional lifestyle of fishing along the coast and farming on Kīlauea's slopes. Several men working in official capacities with the National Park Service in the 1930s observed that the native lifestyle was rapidly disappearing and moved to find a way to preserve the Hawaiians' unique way of life.
Courtesy of Lyman House Memorial Museum.

With the goal of preserving and protecting these last few truly Hawaiian communities, an extension to the national park was proposed in 1932. The Kalapana Extension finally authorized by Congress on June 20, 1938, did not include the lands and community of Kalapana as originally envisioned. Congress had counted on a donation of private lands to be added to the acreage already controlled by the government, but the donations never materialized. Instead, the Hawaiian territorial government paid landowners of Hawaiian ancestry less than one dollar per acre in the acquisition of lands condemned for the extension of the national park.

Although some in the Hawaiian community encouraged the concept of including Kalapana, Kaimū, and Kapa'ahu in the national park as a "living museum," the residents themselves unanimously opposed the proposal. After the loss of many ancestral holdings in the ninety years following the adoption of the western system of private lands, Kalapana's Hawaiians were not about to relinquish their last tenuous claim to the volcanic lands of their birth and sustenance.

FOLLOWING SPREAD:
Kaimū resident Darryl Kealoha casts for fish as his father and grandfather did before him.

CHAPTER 7
Puʻu ʻŌʻō to Kūpaianaha

Episodes of fountaining, occasionally as high as fifteen hundred feet, characterized the first three-and-a-half years of the East Rift Zone eruption. Such fountains built a 835-foot cinder cone, Puʻu ʻŌʻō.

Nineteen eighty-three was only a few hours old when incandescent lava burst through the blackness of the night near Nāpau Crater in Hawaiʻi Volcanoes National Park, then moved downrift in an effervescence of fiery fountains and caustic fumes. Over the next few months, in episodes usually lasting a few hours, intermittent jets of lava—as permeated with gas as bottles of champagne—spewed several hundred feet into the air through migrating lines of fissures.

It took until June for the eruption to localize and begin building a cone. Scientists called the growing cone the O vent, a reference to the vent's plotting on the U.S. Geological Survey map next to a printed "o." The Hawaiians downslope were more poetic: When asked to bestow an official name on the new landmark, they dubbed the cone Puʻu ʻŌʻō in honor of an extinct native bird.

The exotic beauty and novelty of the eruptive episodes, which settled into an almost monthly regularity, fascinated the island's residents and visitors. In addition to the spectacular aerial displays visible from locations throughout the Hilo and Puna districts, the fountaining had audible accompaniment for those within earshot, a thunderous venting that sounded "like the vents were chanting" to some who lived in Kalapana.

But the eruption was not so alluring to those in the immediate path of the lava. The volcano's attack came swiftly. In January, within a week of Kīlauea's brilliant beginning, fast-moving ʻaʻā flows chased from their homes residents of Royal Gardens, a sparsely developed subdivision laid out on the slope just below Kīlauea's ridgeline on lands abutting the national park's Kalapana Extension. Two dwellings were destroyed by March 3, another six were incinerated in April, and between June 29 and July 4, 1983, eight more homes were consumed. While outsiders awaited

53

each new eruptive phase with eager anticipation, Royal Gardens residents learned to live with the possibility that they could lose their homes on less than twenty-four-hours' notice. Mingled with this anxiety was the choking smoke of burning forests and loud explosions as lava hit pockets of methane gas in the thick vegetation, all punctuated by evacuation orders issued at every threat by Hawai'i County Civil Defense.

Despite the losses in Royal Gardens, there was no sense of alarm in the older community along the Puna coast, or in the halls of government at the county or state levels. The Hawaiians whose families had lived at Kapa'ahu, Kalapana, and Kaimū for generations and the residents of other subdivisions nearer the ocean felt no threat from Pele. Occasionally, airborne by-products of the high-fountaining outbursts, sometimes looking and feeling like pieces of dirty styrofoam, other times in the form of cinders and Pele's hair, rained down on the lands between Pāhoa and Pahala. This tephra debris appeared to do minimal harm and added to the fascination of those living on "the Volcano Island."

While the scientists at the Hawaiian Volcano Observatory had forecast the initial outbreak of eruptive activity, throughout 1983 they were hesitant to predict just what the volcano would do next. In February 1984, the Observatory's scientist-in-charge,

In July 1986 the subterranean pipe leading to Pu'u 'Ō'ō was no longer able to contain the forces driving the spectacular high fountains. The ground around the vent ruptured, setting the stage for the formation of a new vent two miles down the East Rift Zone. Lava streamed incessantly from this new outlet and formed a lava shield around a molten pond. After consultation with the elders in Kalapana, the evolving landmark was named Kūpaianaha.

Robert Decker, warned that Kīlauea's outpouring could be "a very long-lived eruption" and he expressed concern that the development of subterranean lava tubes could begin delivering molten lava far from the vent, into populated areas.

The following month, Mauna Loa rumbled to life and for three weeks sent lava flowing toward Hilo. When Kīlauea's episodic fountaining kicked in on its own monthly schedule on March 30, it marked the first simultaneous eruption of the two volcanoes in sixty-five years.

The 1984 Mauna Loa eruption ended without mass destruction. However, less than two weeks after the Mauna Loa activity ceased, continuing flows from Kīlauea's Puʻu ʻŌʻō made the first foray into the traditional homelands of the Hawaiians downslope. Three buildings and a piggery at Kapaʻahu fell to the power of an ʻaʻā flow that stopped half a mile short of Highway 130. The lava snuffed out the farm of Akima Ah Hee, one of the last taro farmers in the area, but the event marked the beginning of a two-and-a-half year respite from destruction. No more homes would be consumed by Kīlauea's fiery fingers until November 1986.

Kīlauea continued its spectacular episodic fountaining on a regular basis throughout 1984, 1985, and into 1986, through forty-seven

This aerial view of the Kalapana coastline, looking south, shows the finger of lava that reached across Highway 130 to incinerate homes in Kalapana Gardens in December 1986. The massive flow field beyond is the Kapaʻahu flow that first reached the sea in November of that year.

Toes of pāhoehoe *lava emerge from the front of an* ʻaʻā *flow*

For the first four years of this on-going eruption, lava repeatedly found its way to the coast, traveling through lava tubes that were at times revealed by collapses referred to as skylights.

eruptive events. In mid-1986 the fountaining, which had occasionally rocketed to fifteen hundred feet above the 835-foot Puʻu ʻŌʻō cone, ceased abruptly and surface activity shifted to a ground fissure nearly two miles downrift. By the end of July 1986, lava was spewing continuously from this new vent, and within weeks the non-stop streaming of molten rock had created a lava lake. The elders in Kalapana began referring to the lake's growing shield-type mound of lava as Kūpaianaha, a Hawaiian word meaning strange, surprising, extraordinary, unaccountable.

During the first four months of continuous activity, the eruption received little public notice. Then the molten lava (two thousand degrees Farenheit or more) spewing from the remote, generally inaccessible vent at Kūpaianaha began forming a tube system beneath the surface, moving stealthily closer and closer to the cluster of communities along the coast. In mid-November, having dropped to an elevation of sixteen hundred feet in a gravitational quest for the sea, the tube system began feeding flow after flow onto the slopes above Kalapana until one flow made it into the populated area. It was a heartbreaking Thanksgiving that year for nine families in Kapaʻahu Homesteads as the lava poured downslope, consuming their homes and everything else in its path. For some, it was everything they owned. By November 26, the flow overran Highway 130. On November 28, lava entered the ocean for the first time since 1973, an act many had hoped would quell the volcanic furies.

But it was not to be. By the end of November 1986, Hawaiʻi County Civil Defense Administrator Harry Kim told the community it was time to "sit down and discuss the long-term problems," the first public notification that the eruption could go on "for days, months, even years."

The threat eased briefly in early December, after lava had ignited a tenth home at Kapaʻahu and consumed more than half a mile of highway, isolating several residences and blocking the coastal entrance to the national park. But the interruption in the lava cascading downslope through the volcano's subterranean tubes proved short-lived. Within two weeks, lobes of molten material moved to the east, reaching into Kalapana for the first time. The fast-moving river of lava emerging was just beginning its rampage. Eighteen residences fell to the lava between December 17 and 20, fifteen of them in Kalapana Gardens subdivision on the southern end of Kalapana. At one point, the lava took ten homes in a hundred minutes. An evacuation notice was issued for the entire area surrounding Kalapana Gardens. Then as quickly as the lava had started pouring into Kalapana, it stopped.

As a modern hula *master and musical composer whose ancestors lived at Kapaʻahu, Johnny Lum Ho keeps the ancient chants and legends of the Kalapana area alive through the dances and songs he shares with the world in performances at Hilo's annual tribute to the* hula, *the Merrie Monarch Festival.*

Pele is the mythical ancient goddess of the volcanoes of Hawaiʻi. With her brothers and sisters, she traveled in spirit from Tahiti to make her home in the tumultuous firepit of Kīlauea which had been burning from time immemorial. When Johnny Lum Ho, a contemporary *hula* master, wrote this chant, Kīlauea's East Rift Zone had been erupting for four years. His mother's birthplace at Kapaʻahu had been spared, however temporarily, from the devastation of a sea-going flow that swept through Kapaʻahu Homesteads, burying under a suffocating blanket of lava the homesteads of islanders who had lived with the volcano for generations.

But Lum Ho is a Christian, and he is quick to tell you that even though he is Hawaiian, there is but one true god and it is not Pele. Likewise, while most people in Hawaiʻi today do not worship Pele, many, especially those occupying her traditional terrain, do express a belief in her as the personification of Kīlauea's awe-inspiring natural phenomena.

Ka Luahine Piha Inaina ʻO Puna	On again, off again, on again, off again
Ua ʻā, ua pio, ua ʻā, ua pio	The old woman that lives in the mountains
Ala hou mai ʻoe	Flowing down from Puʻu ʻŌʻō
E ka luahine no i ke kuahiwi	Threatening all things in its pathway
Kahe mai ana o Puʻu ʻŌʻō	
Ua hoʻoweli nona mea apau	
Ma kona alahele lā	
Hōʻea mai nei, ka ʻāʻā enaena	The red hot *ʻaʻā* is coming
Luaʻi mai ka waha	Vomiting from the jealous woman's mouth
O ka wahine lili (grr-grr)	The munching *pāhoehoe*
Nome mai ka pāhoehoe	The crawling fire destroys
Hailuku ke ahi	
I kokolo mai nei	
Hakukoʻi e ka ʻino	The raging red river flows
Ke kahawai inaina ulaula (grr-grr)	Exploding fire eats the forest
Pahūpahū ke ahi, e ʻai ka nahele lā	The mad and vengeful woman (growls)
Ua huhū ia, mauhala ka wahine	
Ua lilo i ke ahi o ka ʻululāʻau	The forest is burned and lost
Lūluku ka ʻāina o Kapaʻahu	Kapaʻahu is destroyed

63

"Lava was a Latin word for that state of redness that comes out from the mountain; the Hawaiian word for that portion that is hot and moving is pele."

—Piilani Kaawaloa

Waiaka ponds

CHAPTER 8

The Beginning of the End

The winter of 1986 marked the beginning of the end for Kalapana, despite the fact that it would be more than three years before Kīlauea again threatened the little town itself. Throughout 1987, 1988, and 1989, lava mercilessly picked off homes in Kapaʻahu and the Royal Gardens subdivision as the tube system feeding from the lava lake at Kūpaianaha continued to establish its destructive underground path down the volcano's eastern slopes. The serene Waiaka ponds were filled in early in December 1986. Punaluʻu, a brackish pond and popular swimming hole more commonly known as Queen's Bath, disappeared under the encroaching flows early in the spring of 1987. By the end of that year, the tally of dwellings destroyed by the eruption rose to fifty-eight, each residence's distinguishing features, every familiar landmark buried under a suffocating blanket of lava that in some places grew to a depth of seventy-five feet.

It was an agonizing assault at Kapaʻahu, a precursor of what was to follow at Kalapana. Flow after unpredictable flow emerged from the tube system, sometimes in fast-paced fluidity, other times inching forward slowly before entombing everything in its path. It was a pattern that would be repeated innumerable times over the next few years. The erratic nature of the lava's attack played havoc with the psyche of the threatened population. It was extremely difficult for those who had built homes there, worked the land and raised their families there—in many instances for several generations—to realize that their neighbor the volcano had become a very real, very personal threat.

Despite their expressed acceptance that they were living in a land ruled by the volcano and therefore subject to its forces, many seemed to want to deny the reality of the events taking place. In their search for an understanding of the tragedy, some

In the mid-1800s, Father Damien de Veuster built a number of churches in Hawai'i, singlehandedly hauling great logs from the slopes to construct the buildings that helped establish the Catholic Church in the islands. Puna was the first assignment for the Belgian priest who later became known for his heroic dedication to the victims of Hansen's Disease on Moloka'i and who is today a candidate

turned to their religious faith, others invoked the power of Pele. Nevertheless, bewilderment and shock set in as family after family was displaced by the on-again, off-again advance of the lava. It was a nightmare of unpredictable length and meaning, and the uncertainty added to the trauma. Homes just inches from disaster were often spared by a momentary pause in eruptive activity, only to face destruction with the return of new flows days, weeks, or months later.

Throughout 1987, Kapa'ahu bore the brunt of the volcanic onslaught. In 1988, the lava repeatedly overran the reputed site of the first Catholic church built in Hawai'i by Father Damien de Veuster. Father Damien was the stalwart Belgian priest who became known among the Hawaiians for his prodigious strength and, later, for his devotion to those stricken with Hansen's Disease who were cast out of society to suffer in isolation at Kalaupapa, Moloka'i.

As Kīlauea's arteries spewed layer upon layer of lava onto the southern side of Kapa'ahu, an immense field of black rock was created, fanning seaward from the slopes and into the eastern reaches of the national park. Intermittently, the flows began to threaten the archeological remains at Waha'ula Heiau and the National Park Service's Waha'ula Visitor Center. Emergency efforts were launched in a last-ditch attempt to record and preserve any undocumented archeological remains at the Waha'ula complex. Park service fire fighters battled the invading mass but on June 22, 1989, flames consumed the breezy, single-story visitor center despite the fire fighters' best efforts. Three weeks later, the flows ignited the three park rangers' residences adjacent to the visitor center. Lava overran parts of Waha'ula Heiau but the main enclosure of the *heiau,* the walls that surrounded the ancient Hawaiian war temple, was spared.

for canonization. Parishioners at Kalapana's Star of the Sea Catholic Church memorialized Father Damien in paintings and a large wooden carving. Lava repeatedly overran Kapaʻahu coastal areas in 1987 and 1988, encasing the remains of Damien's first mission on property that in the 1980s was developed as Pacific Paradise Estates.

Close-up of lava pouring into the ocean at night

Persistent as man was, Nature was equally persistent. By mid-afternoon on June 22, 1989, lava had crawled all the way under the visitor center. Fire fighters persevered but lost the battle when water ran out. Within minutes, the visitor center began to burn.

82   *Built of steel, the superstructure of the Wahaʻula Visitor Center remains imbedded in the virgin landscape.*

New land was added to the Puna coast after the lava overran the park complex, where it left the visitor center smoldering (above) but spared the ancient heiau. Steam rises (right) as a new beach is created fronting the still-standing thirteenth century "Temple of the Red Mouth."

CHAPTER 9
The Onslaught

As the eruption filled in ground recesses above Highway 130 during the first few months of 1990, disaster officials began warning Kalapana's residents of the buildup of a massive, slow-moving "glacier of lava" getting ready to descend on the town. The growing mass was obscured in daylight hours by the uneven landscape and thick vegetation, but the lava's glow at night signaled the approaching firestorm.

Disregarding the volcanic destruction ongoing throughout the 1980s, people bought land and built new homes in Kalapana, particularly in Kalapana Gardens subdivision, in numbers unprecedented in the sleepy coastal community. Some of the families from Kapa'ahu and Royal Gardens who had been displaced by the volcano relocated to Kalapana and Kaimū, less than five miles up the road. As Kīlauea's outpourings continued southward throughout 1987, 1988, and 1989, a veneer of safety prevailed.

In January 1990, a pause in the eruption at the Kūpaianaha vent drained the lava pond. After a four-day lull, lava once again cascaded downslope from the vent, bypassing a ridge that had built up along the national park side of the flow field over the preceding three years. This time, the flows turned east, toward Kalapana.

In late March, a massive buildup of lava poured into low-lying areas above Highway 130. Although officials had alerted the community to the renewed activity, the threat did not seem real to the residents. Civil defense director Harry Kim estimated that three-fourths of Kūpaianaha's daily lava output of 500,000 cubic meters was filling in the depressions just upslope of Kalapana. But, he said, from the roadway it was hard to see the buildup, even though it was less than two hundred yards away. "The feed into this area was steady, spreading over such a huge area that forward progress was at times immeasurable. We could see an inflation factor building up, but you could not notice this unless you looked at it every single day and you had a measuring device. You're talking literally of a mountain of lava behind the flow front," he explained.

Watching the steady growth of what he described as a glacier of lava, Kim knew by April 1 that the only hope for Kalapana

An agonizing deathwatch began for Kalapana Gardens residents in April 1990 as lava flowed into their yards and then stopped, broke out elsewhere, then returned to flow into their yards again—sometimes under their houses—and then stopped. But it did not always stop, and eighty houses were ignited over the next two months.

"What made Kalapana Gardens so special for us was all the loving people there. I had never been anywhere where I felt so welcome and comfortable immediately."

—Mary Dressler, Kalapana Gardens resident

Gardens was a halt in the eruption or a disruption in the tube system feeding the flow. "Once it crossed that highway," the civil defense administrator said, "we knew we would have to issue evacuation orders."

The hoped-for disruption in the eruptive activity failed to materialize. After several weeks of a very slow-moving flow front, lava was poised above Kalapana, ready to descend at the slightest gravitational nudge.

"By April we were already telling people 'You have to plan for evacuation, you have to plan for the possibility of inundation and, unfortunately, you have to plan for permanent relocation,'" Kim said.

As lava ate its way across Highway 130 between Keone Road and Kalapana Avenue on April 2, 1990, the evacuation order was given to several dozen Kalapana Gardens residents. At 6 A.M. on April 4, the lava torched a house belonging to Ruth Duff who, with her late husband, Jack, had bought six hundred lots in the 1960s and created Kalapana Gardens. Five hours later a second Lokelani Street home fell victim to the flow.

Suddenly the eruption shut down, allowing the flow into Kalapana Gardens to crust over, but the pause lasted only a day.

A lobe of lava threatening the community begins to fill the back end of the brackish pond abutting Hakuma Point.

the church for their children, the issue even dividing some families. Finally, the Catholic diocese in Honolulu stepped in and ordered the removal of the church. Once the decision was made, a band of men, some parishioners and some members of the island community outside the Kalapana area, united in a single goal: to save the colorful seaside church.

As April drew to a close, the flows, now moving on three separate fronts, seemed to slow, but as the lava inched its way into the village itself, tons of molten rock were flowing into the back areas of Kalapana Gardens, mercilessly picking off home after home that had been spared in earlier surges. The stall at the front bought time for those who were packing up the two churches' belongings. While the contractor, Mike Latimore, and his crew worked to brace the walls and windows, both inside and out, of the Painted Church, the volcano was gathering its forces for a new assault.

On May 2, with huge volcanic steam clouds looming overhead, civil defense authorities told the local general store proprietor, Walter Yamaguchi, it was time to close his Kalapana Store and Drive Inn. The following day was to be the final opportunity for those remaining in the area to remove goods and personal belongings in advance of the evacuation of the Painted Church. But as evening descended on May 3, a sudden, massive influx of lava began pumping into the flow front. Overnight, a fiery stream of lava burned a path behind Yamaguchi's store and between the Congregational Church and the three homes behind it. Suddenly, Beach Road, the church's escape route, was threatened.

During the first months of the disaster, Hawai'i County Civil Defense established a local command post (right) at Harry K. Brown Park. As the eruption progressed, the displaced members of the community also sought refuge there and a tent city grew, albeit temporarily, as eventually the lava found its way to the park, too.

Throughout the night of May 3, in an impressive display of intensity of purpose, all energies were directed toward securing the 28-by-59-foot church structure for the final move. At dawn the following day, the trailer was backed underneath the building. In a race against the fast-moving flow advancing just across the street, the workers maneuvered the Painted Church the short distance down Beach Road to Harry K. Brown Park, finishing just hours before the lava cut off the last paved access into Kalapana.

Simultaneously, lava began filling the brackish ponds beneath Hakuma Point immediately adjacent to the recently vacated site of the Star of the Sea Church. But most of the lava was feeding into the lobe that had cut Beach Road and was pushing toward the sea at the popular surfing beach fronting the park. Now it was Harry K. Brown Park, with its ancient stones, the ruins of a *heiau*, and Wai'ākōlea Pond, that was threatened.

Since the onset of the volcano's attack on Kalapana in early April, the county's historic park had served as the command post for civil defense authorities, the phalanx of county, state, and federal workers providing disaster relief for the refugees, and the media. The park's large pavilion had been the scene of numerous community meetings convened by Civil Defense Administrator Harry Kim to keep the residents apprised of the latest developments regarding the eruption. Evacuation notices and other important communiques were posted on a bulletin board at the park and the Red Cross provided meals for the evacuees and disaster relief workers from a mobile lunch wagon there.

Nearly a month elapsed between the time of the lava's first foray into Harry K. Brown Park

and the flood evident in the aerial view above.

The glossy sheen of the three-pronged flow on May 7, 1990, was in stark contrast to the lush vegetation and remaining beach community.

CHAPTER 11
The Finality

*"It's God's will;
it's not ours."*

—Bernice McKeague,
Kalapana Mauna Kea
Congregational Church

It was a lingering death for Kalapana. The isolated hub of the town remained standing for nearly a month as another pause in the volcanic outpouring followed the firestorm of April and early May. In the latter part of May, the volcano resumed eruptive activity, issuing wave upon wave of fluid lava which, solidifying as it cooled, stacked layer upon steaming layer of *pāhoehoe*. By summertime Kalapana was buried under fifty to seventy-five feet of fresh, hardening lava. One by one, the remaining homes in Kalapana's old and new communities went up in flames. On June 1, as midnight approached, the ever-tightening circle of lava set the humble Congregational church ablaze while four members of its congregation watched. The former site of the Star of the Sea Painted Church was next. The Catholic church's recreational building was consumed June 3 as the lava made its way to the ocean. And on June 6, Walter Yamaguchi's defiance of the volcanic odds met with defeat as the Kalapana Store and Drive Inn succumbed.

All that then remained of Kalapana was approximately a dozen homes scattered around the edges of the myriad flows. By summer's end only a handful of residences remained, and in its relentless cycle of destruction and construction, the volcanic steamroller had pushed eastward into Kaimū, reclaiming the world-famous Black Sand Beach as waves of lava filled in Kaimū Bay and thrust the island's coastline more than half a mile seaward.

Society has no means of calculating the true losses sustained by the communities of Kalapana, Kaimū, and Kapaʻahu as a result of Kīlauea's eruption at Puʻu ʻŌʻō and Kūpaianaha. In its inundation since 1983 of an area of approximately thirty square miles, the eruption destroyed 175 residences and forced the removal of

twenty-one uninsured homes, as well as the Painted Church, made more than two dozen other homes uninhabitable, and covered more than one thousand residential lots. Some property owners were still paying for land that became buried under tons of lava, land which lost any real market value while property values elsewhere on the island were booming. A handful of businesses in Kalapana were directly affected by the eruption, and airborne volcanic pollutants had a detrimental impact on agricultural interests islandwide.

The task of putting a price tag on the damage done to private and public properties, facilities, roads, and utilities fell to the county and state governments. By May 1990, they had tabulated estimates of nearly $50 million in damages to private property, and another $15 million to public facilities. The cost of replacing the public and private roadways covered by the lava was put at more than $32 million; nearly $3.5 million would be needed to rebuild the water systems.

Impossible to reconcile on any accountant's ledger are the priceless intangibles—the unrecorded history of the homesteads

Lava finds its way into the ocean after setting the Star of the Sea's recreation building on fire (seen in the background). While government agencies tabulated the finan-

cial toll of the federally declared disaster, the loss to Kalapana's residents of such intangibles as access to the sea was immeasurable.

and the community that nurtured and sustained generations of islanders, the ancestral graveyards, the access to lands and oceanfront from Kamoamoa to Kaimū, the groves of fruit trees and beautiful flowering plants cultivated throughout the area, and the hopes and dreams of all who called Kalapana home.

The state and federal declarations of disaster made the refugees eligible for housing relocation assistance, tax relief, and low-cost loans. A task force made up of representatives from a variety of government agencies was convened to coordinate relocation efforts while bills aimed at providing relief were being considered by state legislators.

Armed with a U.S. Geological Survey assessment of volcanic hazards that essentially said no place in lower Puna was safe from deluge by lava, the Federal Emergency Management Agency recommended to the state and county that resettlement of Kalapana be discouraged, as well as any future development of other areas beneath Kīlauea's East Rift Zone, a recommendation bolstered by private insurance companies' refusal to insure such properties.

Meanwhile, the federal government was providing temporary housing assistance to approximately fifty families and another twenty families were being helped in their relocation efforts by the American Red Cross. Some of the families dispossessed by the 1986–87 destruction at Kapaʻahu had already rebuilt after accepting developer David Watumull's offer of one-acre lots in the Puna subdivision of Hawaiian Paradise Park. Although most of the displaced residents stayed at least temporarily in East Hawaiʻi, some of the Kalapana residents moved off-island, some relocating to the Mainland, some to foreign countries, according to Harry Kim of the civil defense.

"There is that percentage who cannot bear the thought of going back, who cannot accept a compromise between what they had and what may be developed for them." Kim said. "Those are the ones who said 'I'm going; I'm not coming back,' who never want to expose themselves to such trauma again. And they have been told, as long as they stay in the Kalapana area, that's a possibility."

There were others who, a year after the destruction, were still in a state of limbo. David Riddley, Hawaiʻi County Mental Health administrator, said many of the Kalapana residents suffered from a post-traumatic stress disorder similar to that experienced by veterans of war. The drawn-out nature of the catastrophe put the residents through a torturously prolonged period of uncertainty, not unlike that faced by those living in a war zone. The hot lava incinerated and then obliterated almost everything, leaving many unable to even identify their former homesites in the new landscape. With the exception of a few sheets of metal roofing and the burned-out hulks of discarded vehicles captured in the fresh rock, the volcanic destruction left nothing for the residents to pick up in order to go on.

But there are those who want to return to the area and build anew. Those expressing the strongest desire to return to the land are Kalapana's native Hawaiians.

As the agonizingly slow process of lava inundation continued, the remains of the once pristine and fondly remembered coastal community littered the imposing new volcanic landscape.

Having bypassed the Kalapana Store and Drive Inn and overrunning the grounds of Kalapana's two churches, the lava entered the ocean at the historic canoe-landing site.

Since 1977, Walter Yamaguchi, at right, had disputed geologists' predictions of imminent demolition by lava, saying he felt Pele would protect his Kalapana Store and Drive Inn. Yamaguchi's store remained untouched when almost all other developed property had succumbed to the volcanic onslaught. The store finally fell to the unyielding tide on June 6, 1990.

When Pāhoa businessman Walter Yamaguchi bought property in Kalapana in the 1930s, his family and friends, even his mother, thought he was crazy. And they figured he was just throwing more good money after bad when he built a store on the property in 1974. Walter Yamaguchi says if he was crazy, he was crazy like a fox.

The octogenarian storekeeper recalled the history of his store and its fateful end in the 1990 lava flow. "Harry Kim was telling me I had to get out. I didn't want to go. But I made $6,000 CASH that last day."

Walter and Maizie Yamaguchi's Kalapana Store and Drive Inn was billed as the "oldest water well in Puna." The little hamburger stand and adjoining store, built of cinder block and corrugated roofing and powered by a gasoline-run generator, was the hub of the commercial, and often the social, activity in the quiet coastal village. The wall fronting the store displayed a bulletin board of local events and notices of surfboards or cars for sale, lost dog reports, and the other incidental messages of the laid-back surfing and retirement community that relied on what was commonly referred to as "Walter's store." The Kalapana Drive Inn was the place the old-timers, county workers, neighbors, or those who had come to pick up their mail from the bank of stainless steel post office boxes outside liked to sit for a cup of coffee and a jocular discussion of the latest political brouhaha being dished up in the daily paper. The pay phones outside were in frequent use, with people queueing up to make calls during breaking news events. The two public phones were the only contact points with the outside world for families displaced during the long siege, as well as for the journalists who were dispatched to cover the events. Like the celebrated Hasegawa General Store in Hana, Maui, you could get everything at Walter's store—from snacks, ice, and your favorite six-pack to wicks for kerosene lanterns, fishhooks, and zoris for reef-walking.

But it was Yamaguchi's seemingly insurmountable belief in 1990 that

As if beseeching the volcanic forces to stop, a road block marks the end of the road at Kaimū Bay in this nighttime shot. Neighbors wondered at the erratic nature of the lava that took everything from one Kaimū landowner while leaving the property next door intact.

This aerial photograph, taken on December 7, 1990, from twenty thousand feet above sea level, shows that after engulfing Kalapana, the lava found its way back to the western side of the flow field, now more than five miles wide. A large plume of gases and water vapor is evident, rising from the vent of Puʻu ʻŌʻō. Mauna Loa and Hualālai are seen in the background, and the lava is seen flowing into the ocean near Wahaʻula.

CHAPTER 12
The Legacy

Aloha 'āina, a love of the land as well as the forces that created it, has been the Hawaiians' strength for more than fifteen hundred years. After centuries of living with the volcano's titanic powers, the inhabitants of Kīlauea's southern slopes find themselves in a different struggle for survival in the twentieth century.

Piilani Kaawaloa was raised in the Hawaiians' time-honored system of *hānai*. Her maternal grandparents, William and Minnie Kaawaloa, whose Kalapana home occupied an enviable site on the oceanfront outcropping known as the Hakuma horst, took Piilani into their home to nurture and care for her as their own child. To Piilani, her upbringing at the hands of her grandparents was a blessing, because they practiced many of the traditional ways of their ancestors and retained knowledge of native skills, expertise that has been almost totally lost in the islands.

Well into the twentieth century, Kalapana's isolation allowed the community there to hold on to its indigenous ways, with many of the elders continuing to speak in their native tongue to the present day. But after World War II, the tide of outside influences began to seriously erode the traditional lifestyle.

The Kaawaloas were among those who had opposed pressures to bring electrical power and resort development to Kalapana. Other local families, however, wanted the modern conveniences of electricity, and many felt hotel development would provide jobs so their children would not have to leave Kalapana. The 1977 eruption ended talk of resort development as the county administration, based on information provided by the U.S. Geological Survey, subsequently eliminated all resort zoning in the Kalapana area.

Early in the 1980s, the national park's Chain of Craters Road was connected to a new highway along the coast. Electrical power lines came to Kalapana soon thereafter. And visitors to the insulated cluster of coastal communities came by the hundreds, then thousands. With more visitors came more settlers.

Soon Kalapana's identity as a place of traditional Hawaiian values was overshadowed by a new community of outsiders. The elders complained that Kalapana's young people were losing the ways of old and no longer knew how to fish and make the land

productive, that they cared more about television, automobiles, and the attractions of city life than about maintaining the Hawaiian lifestyle.

When Kīlauea's long-running eruption began in 1983, Piilani Kaawaloa was pursuing Hawaiian studies and a degree in education at the University of Hawai'i at Hilo. As the volcanic outpouring continued and eventually consumed family members' homes and buried beloved features of the landscape, her emotions ranged from sadness to anger—sadness at the inestimable losses, and anger at the influences of outsiders that had and continued to interfere with the old-style life of Kalapana. Perhaps, she thought, Kīlauea's destructive activity was a predestined cleansing of the land.

"I feel sorry for our people. They can't afford to buy land any more. The only ones who can afford to buy land are the ones who go off to the Mainland or somewhere else to work to make big money. Why is that? Why do we have to move away from our own island to make money to try to outbid the person who just came in? At the same time we're fighting for our land, we're fighting for our culture. We're losing it.

"Here we are, Hawaiians in our own land, but most of us can't even speak our own language," she lamented. "Our language is different and unique from the other Polynesians. We cannot go back to some place and find our language or be retaught. All our grandparents are dying off and they are the only ones who can teach us."

As Kaawaloa went about her routine of teaching in the pub-

Piilani Kaawaloa speaks of following the ways of her ancestors who colonized these lands. She sees opportunity in the new lands created by Kīlauea for the Hawaiians from Kalapana who can no longer afford to buy land.

As if answering the call of all the generations who have come before her, ipu and hala shoots in hand, Kaawaloa makes a pilgrimage across the fresh landscape to a new beach that formed along the extended coast at Kaimū.

lic schools and studying at the university in 1990, her anger began to soften. She began to envision an opportunity for her people to revive their culture: Kalapana now represented a new land. Her ancestors had traveled thousands of miles to settle new lands. It would be hard, it would take a different kind of pioneering effort, but there was much to gain and nothing to lose. The effort could begin with preparation of the land; she could begin by cultivating seedlings of the plants that had sustained the ancient Hawaiians.

Throughout the latter part of 1990 and into 1991, the lava took the western track, flowing into the sea at points south of Kapaʻahu and within the boundaries of Hawaiʻi Volcanoes National Park. By the spring of 1991 there were already signs of new life at Kaimū. Hawaiians living nearby had already begun planting coconuts and other vegetation in the virgin sands of a beach that had formed along the extended Kaimū coast.

So as the federal and state bureaucrats wrestle with the legalities of allowing the Kalapanans to return to their land, Kaawaloa puts her energies into perpetuating her culture. In the language and dance classes she gives several times each week, the energetic young woman freely shares the precious legacy bestowed upon her. She likes to focus on the youngest members of her society, immersing them in the language while trying to instill in them pride in their Hawaiian heritage. She encourages her older students too. "You can still learn the language. It may be changing a little, but at least we're still speaking Hawaiian."

Ua mau ke ola ʻo Puna. Aloha ka ʻāina ʻo Kalapana.

Forever Kalapana.

Hala *and coconut, for which the Hawaiians had many uses, were among the seedlings carried from the far reaches of the Pacific to sustain the early Polynesians. Kalapana and the Puna coast were known throughout antiquity for an abundance of both.*

"I see that there are a lot of things that we can do only because they have declared this a disaster area. Nobody can build a hotel, nobody can build anything sophisticated," says Piilani Kaawaloa. "Why couldn't we just make it our home again?"

"Ua 'imihia ka māno wai
o ko Hawai'i kūpuna."

A Hawaiian proverb meaning:

*"The many sources of waters
and life of the ancestors of
Hawai'i have been sought after."*

Bibliography

Allen, Melinda Sue. 1979. "The Kalapana Extension in the 1800s: A Research of the Historical Records." National Park Service, Hawai'i Volcanoes National Park. Typescript.

Apple, Russell Anderson. 1954. "A History of Land Acquisition for Hawaii National Park to Dec. 31, 1950." Masters thesis, University of Hawaii.

———. 1971. "Homesite Provisions of the 1938 Kalapana Act." National Park Service, Hawai'i Volcanoes National Park. Typescript.

Baldwin, Helen Shiras. 1970. "April Anniversary Month of Great Quake, Disaster." *Hawaii Tribune-Herald Orchid Isle* April 12, p. 4.

Ballard, Robert D. 1983. *Exploring Our Living Planet*. Washington D.C.: National Geographic Society.

Barrère, Dorothy B. 1962. "National Survey of Historic Sites and Buildings Theme XVI: Indigenous Peoples and Cultures. Hawaii Aboriginal Culture A.D. 750–A.D. 1778." U.S. Dept. of Interior, National Park Service. Hawai'i Volcanoes National Park. Typescript.

Beckwith, Martha. 1971. *Hawaiian Mythology*. Honolulu: University of Hawaii Press.

Bostwick, Jr., Burdette E., and Brian Murton, editors. 1971. "Puna Studies: Preliminary Research in Human Ecology 1971." Committee on Human Ecology, University of Hawaii. Typescript.

Brigham, William T. 1909. "The Volcanoes of Kilauea and Mauna Loa on the Island of Hawaii." In *Memoirs of the Bernice Pauahi Bishop Museum,* vol. 2, no. 4. Honolulu: Bishop Museum Press.

Brumaghim, Everett. 1933. Letter to E. P. Leavitt, superintendent Hawaii National Park. Hawai'i Volcanoes National Park Library.

———. 1933. "Report of Heiau Sites, District of Puna." Ethnological Pamphlet No. 22. Hilo, Hi.: Hawai'i Volcanoes National Park.

Bunson, Margaret R. 1977. *Faith in Paradise: A Century and a Half of the Roman Catholic Church in Hawaii*. Boston: St. Paul Editions.

———. 1989. *Father Damien: The Man and His Era*. Huntington, Ind.: Our Sunday Visitor Publishing Division.

Ching, Francis K. W., Catherine Stauder, and Stephen L. Palama. 1974. "The Archaeology of Puna, Hawaii: Archaeological Surface Survey for the Proposed County Beach Park Improvements." County of Hawai'i. Typescript.

Daws, Gavan. 1973. *Holy Man: Father Damien of Molokai*. New York: Harper and Row.

———. 1974. *Shoal of Time: A History of the Hawaiian Islands*. Honolulu: University of Hawaii Press.

Decker, Robert, and Barbara Decker. 1989. *Volcanoes*. New York: W. H. Freeman and Co.

Ellis, William. [London 1827] [Hawaii 1917] 1963. *Journal of William Ellis: Narrative of a Tour of Hawaii*. Reprint. Honolulu: Advertiser Publishing Co. Ltd.

Emerson, Nathaniel B. 1986. *Unwritten Literature of Hawaii: The Sacred Songs of the Hula*. Rutland, Vt.: Charles E. Tuttle Co.

Emory, Kenneth P., J. Halley Cox, William J. Bonk, Yoshiko H. Sinoto, and Dorothy B. Barrère, 1959. "Natural and Cultural History Report on the Kalapana Extension of the Hawaii National Park, Vol. I, Cultural History Report." National Park Service. Typescript.

Emory, Kenneth P., Edmund J. Ladd, and Lloyd J. Soehren. 1965. "The Archaeological Resources of Hawaii Volcanoes National Park, Part II: Additional Sites, Test Evacuations and Petroglyphs." Dept. of Anthropology, Bishop Museum. Typescript.

Harby, Bill. June 1990. "Harry Kim: Harry versus the Volcano." *Honolulu Magazine,* vol. XXIV 12:46–55.

Hawaii Natural History Association. "Wahaʻula Heiau: Temple of the Red Mouth." Hawaiʻi Natural History Association and National Park Service. Brochure.

Hawaii Tribune-Herald. 1968, 1970, 1972, 1975, 1977–78, 1980, 1983–1990.

Hawaiian Volcano Observatory. 1987. *Hawaii Symposium on How Volcanoes Work: Diamond Jubilee (1912–1987), abstract volume.* Hilo, Hi.: U.S. Geological Survey.

Heliker, Christina, and Dorian Weisel. 1990. *Kilauea: The Newest Land on Earth.* Honolulu: Bishop Museum Press.

Heliker, Christina, J. D. Griggs, Taeko Jane Takahashi, and Thomas L. Wright. 1986. *Earthquakes and Volcanoes: Volcano Monitoring at the U.S. Geological Survey's Hawaiian Volcano Observatory,* vol. 18, no. 1.

Hitchcock, C. H. 1912. "The Hawaiian Earthquakes of 1868." *Bulletin of the Seismological Society of America,* vol. 2:180–90.

Honolulu Advertiser. 1990.

Honolulu Star-Bulletin Advertiser. 1978, 1990.

Ii, John Papa. [1866–70] Articles in *Kuokoa.* In *Fragments of Hawaiian History,* trans. by Mary Kawena Pukui. 1973. Honolulu: Bishop Museum Press.

Kalapana: Death of a Hawaiian Village. 1990. Honolulu: Dobovan Productions. Videotape.

Kamakau, Samuel Manaiakalani. 1964. *Ka Poʻe Kahiko (The People of Old),* trans. by Mary Kawena Pukui, arranged and ed. by Dorothy B. Barrère. Honolulu: Bishop Museum Press.

Langlas, Charles, and Kupuna. 1990. "The People of Kalapana, 1823–1950." Kalapana Oral History Project, University of Hawaii-Hilo. Typescript.

Lyman, Frederick S. 1979. *The Lymans of Hilo: A Fascinating Account of Life in 19th Century Hawaii.* Revised edition. Hilo, Hi.: Lyman House Memorial Museum.

Macdonald, Gordon A., Agatin T. Abbott, and Frank L. Peterson. 1983. *Volcanoes in the Sea: The Geology of Hawaii.* 2nd edition. Honolulu: University of Hawaii Press.

Malo, David. [1898. Trans. by Nathaniel B. Emerson.] 1951. *Hawaiian Antiquities.* 2nd edition. Bishop Museum Special Publication 2. Honolulu: Bishop Museum Press.

Otaguro, Janice. January 1991. "Islander of the Year: Pele." *Honolulu Magazine,* vol. XXV 7:42–55.

Pukui, Mary Kawena, and Samuel H. Elbert. 1986. *Hawaiian Dictionary.* Honolulu: University of Hawaii Press.

Pukui, Mary Kawena, Samuel H. Elbert, and Esther T. Mookini. 1989. *Place Names of Hawaii.* Honolulu: University of Hawaii Press.

Relocation Housing Committee. 1990. "Relocation Housing Plan for Kalapana Disaster Area Victims." Kalapana Task Force. Typescript.

Schoofs, Robert, SS.CC. 1978. *Pioneers of the Faith: The History of the Catholic Mission in Hawaii (1827–1940).* Edited and published by Louis Boeynaems, SS.CC. Honolulu.

State and Federal Hazard Mitigation Team. 1990. "Hazard Mitigation Team Report for the Kilauea Volcano Eruption, Hawaii County, Hawaii." Federal Emergency Management Agency.

Thrum, Thomas G. 1906, 1907. *Tales from the Temples: A preliminary paper in the study of the heiau of Hawaii with plans of the principal ones of Kauai and Oahu.* Part I, p. 49–69. Part II, p. 48–78.

Tilling, Robert I. 1983. "Monitoring Active Volcanoes." U.S. Geological Survey, Dept. of Interior. Pamphlet.

U.S. Geological Survey. 1987. *Volcanism in Hawaii, Vol. I.* Professional Paper 1350. U.S. Dept. of Interior.

Wood, Harry O. 1914. "On the Earthquakes of 1868 in Hawaii." *Bulletin of the Seismological Society of America,* vol. 4 4:200–2.

DORIAN WEISEL conceived and photographed *Aloha O Kalapana*. A resident of the island of Hawaii for the last ten years and of Kīlauea itself for the last five, he operates a stock photo agency, Volcanic Resources, and records volcanic events on film for the U.S. Geological Survey's Hawaiian Volcano Observatory. His first book, *Kīlauea: The Newest Land on Earth* (1990), included photographs which toured with the Smithsonian exhibit "Inside Active Volcanoes." A noted freelance photographer, Weisel has exhibited his work widely, and has published photographs in a variety of publications.

FRANKIE STAPLETON is a staff reporter at the *Hawaii Tribune-Herald* in Hilo. She began her journalism career at the *Daily Press* in Newport News, Virginia. She moved to Honolulu in 1970, where she worked as a television news producer at KITV and completed a B.A. in journalism at the University of Hawaii at Manoa in 1973. Her interest in Kalapana dates back to 1977, when she was assigned as a new member of the *Tribune-Herald* staff to cover the eruption of Kīlauea's East Rift Zone, which sent lava flows dangerously close to the coastal village. Stapleton lives in Puna, not far from the lava flows that now cover Kalapana.